Prop of Materials

Marcia S. Freeman

Contents

What Is It Made Of?

Would you choose the same **materials** to make a birdhouse and a globe?

You might choose wood, nails, and paint to build a birdhouse.

You might choose paper, paint, and glue to make a globe.

Choosing the right materials helps you build or make what you want.

Different materials have different **properties**. Some materials are **transparent,** which means you can see through them.

Other materials are **opaque.** You cannot see through them.

Some materials are rough and scratchy. Others are smooth and shiny.

These are properties of materials that you can learn about by looking or touching.

Testing Materials

You can also learn about the properties of materials by testing them.

You can test to find out if a material is **flexible,** meaning it bends but does not break easily.

You can test to see if a material is **absorbent.** Absorbent materials soak up water and other liquids.

Changing Materials

You can change the way many materials look or feel. You can cut, stretch, shape, or heat them.

Different materials change in different ways. If glass is heated, it will melt. Wood does not melt—it burns.

Choosing Materials

When you know the properties of materials, you can choose the best materials for a job.

What kinds of materials work best for winter clothing?

What kinds of materials work best for a dancer's costume?

What properties do firefighters' uniforms need to have?

Different materials are used to make **packaging,** or containers, for food. Food can come in paper, plastic, cardboard, or metal containers.

What properties does a milk container need to have?

What properties does an egg carton need to have?

What kinds of foods come in metal containers? What materials could you use to package your favorite foods?

Using Many Materials

Many things are made from more than one material. What materials were used to make the parts of this bicycle?

A playground is built of many different materials, too. What materials were used to make this playground?

What materials would work well for a slide? Why?

Houses can be built of stone, brick, or wood. Different materials will work best for the windows, the roof, and the chimney.

What properties do the materials in this tent have? Would you use paper, cardboard, or wood to make a tent?

You use all sorts of materials every day.

Look at the different objects in this kitchen. What materials are they made of? How do their properties compare?

Object	Material	Properties
cutting board	wood	hard opaque
window	glass	smooth transparent
cooking pot	metal	strong
apron	cloth	soft

Glossary

absorbent (uhb-SOR-bunt): able to soak up or
absorb liquid

flexible (FLEK-suh-bul): able to bend or stretch
without breaking

material (muh-TEER-ee-ul): what is used to make
or build things

opaque (oh-PAYK): not clear or see-through

packaging (PA-kij-ing): the container, box, or
other wrapping that an object is sold in

property (PRAH-pur-tee): what makes a material
good for a particular purpose

transparent (trans-PAIR-unt): clear or see-through